Other books in the series:
*The Green Office*
*The Green Garden*
*Green Travel and Transport*

# Green Building and Renovating

## How to live well, be green and make a difference

Tanya Ha

MELBOURNE UNIVERSITY PRESS

MELBOURNE UNIVERSITY PRESS
An imprint of Melbourne University Publishing Limited
187 Grattan Street, Carlton, Victoria 3053, Australia
mup-info@unimelb.edu.au
www.mup.com.au

First published 2007


Designed and typeset by Alice Graphics
Printed by Griffin Press, SA

National Library of Australia Cataloguing-in-Publication entry

Ha, Tanya, 1972– .

Green building and renovating.

ISBN 9780522854466 (pbk.).

1. Ecological houses. 2. Dwellings — Remodelling — Environmental aspects. 3. Sustainable buildings — Design and construction. 4. Environmental protection — Citizenparticipation. I. Title.

690.8

# Contents

# Introduction

Have you ever seen a news story about thousands of hectares of rainforest being felled each year to provide grazing land for hamburger cattle? Have you heard that we're seeing the extinction of species at the fastest rate ever, that the air is filling with greenhouse gases and turning the earth into a toaster oven, and that we're playing Russian roulette with nuclear energy? You probably found yourself caring, but couldn't imagine how you could stop any of it happening. You couldn't see yourself going on a hunger strike or being chained to a bulldozer in protest.

You're not alone. While there are a few hard-core greenies at the frontline of the environmental movement who are not afraid to further the green cause through extreme and very public acts, this approach is not for everyone.

Public protest was (and still is) necessary. Someone had to bring the plight of the environment to the attention of the average person. It took the very public activities of Greenpeace, a hole in the ozone layer, an inevitable backlash against eighties consumerism and a rather gorgeous looking rock star named Sting to make us understand that human activities haven't been the best thing for all creatures great and small.

The truth is that birds are dying because they're covered with oil, ships carrying nuclear waste are probably trying to dock at your local port, and some Pacific islands may well be flooded by rising ocean levels. It's not pretty and it's all our fault!

But I've got good news for the armchair spectator of green sports—protesting isn't the only way to help save the planet. Coming up behind the frontline there's a new wave in the environmental movement, defining itself by what it's *for* rather than what it's *against*. The new environmentalist knows that it's not just the fault of governments and big businesses, and that we all share the blame for damaging the planet. After all, we purchased the products that harmed the earth through their manufacture or disposal. We bought and used the hairsprays that contained ozone-destroying chlorofluorocarbons (CFCs). Perhaps we even wore fur and ate shark's fin soup.

Our generation has made many environmental mistakes innocently. The way to move forward is to take back responsibility for looking after the planet—to stop complaining—and to start concentrating on being part of the solution, rather than just feeling guilty.

So who is part of this new wave of environmentalism? It's hard to tell because they look like ordinary people. They are ordinary people. They have jobs, friends, families and busy lives. They want to enjoy the best and most comfortable lifestyle possible. They're also determined to help save the planet but don't want to waste time just sitting around and talking about it. They would rather get up and do something, by changing their everyday lives, by not using the products that cause harm and by supporting those that are greener. They draw the line at excessive consumption. Their clothes may not be made from hemp, but they'll wash them with a greener detergent in an energy- and water-efficient washing machine, powered by green electricity from GreenPower-approved sources.

They're not heroes, ready to scale tall buildings. They're not selfless saints, making sacrifices for their cause. Far from that, they're actually sensible and slightly selfish people, who happen to be very far-sighted and want to preserve this beautiful planet and enjoy it for years to come. They want enough resources for their children and grandchildren.

Saving the planet starts at home. If you show some care for the planet when planning or choosing and adapting your house, you will be repaid with a comfortable, warm place in which to relax, with lower energy and water bills, and with a cleaner, healthier environment to live in.

Tens of thousands of new houses are built each year in addition to or replacing the millions of existing dwellings. If you're one of the thousands who are building a new house, then you have a unique chance to build a greener, more energy-efficient one that will take a lesser toll on the environment throughout its life. Good design features, such as the right orientation of the house on the block in relation to the sun, can make more of a difference to our energy use than the combined benefits of smaller energy-saving tips, such as putting lids on saucepans or turning off appliances on stand-by. Software has been developed that makes it easy for builders and architects to estimate how energy-efficient a new house is. There are also programs that train and accredit energy-smart and environmentally sensitive builders. It would be great to gradually see eco-homes become more common and the energy-efficient 'green house' become the normal way to build, rather than the exception.

This chapter is a snapshot of what to look for and what to ask your builder or architect to consider. You will need to seek expert advice because individual sites, settings and climate zones have particular considerations, and different local laws and council regulations will have some influence over what you can build in your area and how easy it is to get planning approvals. If you're building or renovating you'll have to make some big decisions, but remember that there are many design features, sustainable materials and new technologies that are readily available. If you understand the issues, you will know what questions to ask.

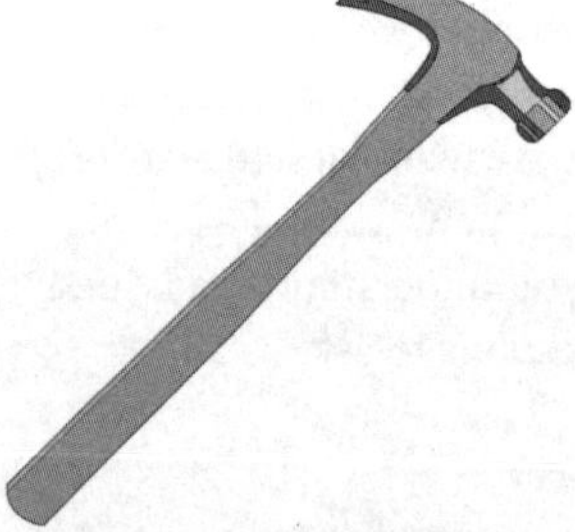

# What makes a house green?

There are new green houses that we tend not to notice because they don't always stand out and they don't fit the hippie stereotype. In fact, a green house can look exactly like any other house. Green features, such as solar panels on the roof or rainwater tanks out in the backyard, are not all that noticeable.

The stereotypical hippie green house and the modern eco-home are both built around the same underlying principles of what makes a house greener.

Key factors in greener buildings are:

- energy efficiency
- water efficiency
- material use.

***Energy efficiency*** is about how well your house uses energy. It involves the appliances, heating, cooling, ventilation and lighting systems of your home. The design of the house and the use it makes of natural light and heat (passive solar heating), natural ventilation and insulation will also determine how much extra heating, cooling and lighting it will need. The term 'energy-efficient house' sounds quite practical and bland, so instead think of energy-efficient homes as comfortable, warm homes that are not expensive to run.

***Water efficiency*** involves water supplies (from water mains and rainwater), how water is used in the home and how wastewater is disposed of.

***Material use*** considers both the sources of the materials used to build your house and whether or not they are sustainable, how they are made or processed before being used in the house, and their effect on how healthy and how efficient the house is to live in.

Looking at buildings from a different perspective, a green house doesn't necessarily make a green home. A green home is a sustainable house in action, fitted and lived in in an eco-wise way. The three factors that make a green home are:

- the house structure—incorporating aspects such as the design and materials used
- fittings—including window coverings and appliances
- occupants—the attitudes and habits of the people who live there.

The occupants are the human element of a green home. Just as a hammer can be a tool in the hands of one person and a weapon in the hands of another, a green house can still have a huge impact on the environment if the people who live there have wasteful habits and ignorant attitudes.

Remember to adapt your lifestyle, as well as your house design.

With green houses, there are many shades of grey. A builder can achieve a house that goes a good way towards being better environmentally by putting in a bit of effort and sometimes a little extra expense up-front. Remember that energy-efficient houses pay back some of their costs through the energy saved over their life span. Enthusiasts can go to more effort and make their houses even greener, but this can mean extra hassle and paperwork in planning approvals, more research and sometimes greater expense. Green house design is not a case of all or nothing. Aim to make your home as green as possible, within your means and the potential of the house or block of land. Think of the greening of your house as an ongoing project. Over time, replace outdated appliances with new, more efficient models, consider the environment when you're making renovations and, if possible, incorporate alternative water and energy technologies into your home.

## Energy ratings for houses

The Energy Rating labels for appliances help you to choose appliances, such as refrigerators or washing machines that need less electricity to run. The fuel consumption label similarly helps you to choose cars that are more fuel efficient, meaning less money spent at the petrol station. Rating schemes allow you to compare the environmental performance of similar products.

There are now a number of rating tools (software, rules and/or mathematical formulae that calculate a number that represents a building's environmental performance) that give new or existing houses a rating—a standard measure of a given house's performance, sustainability or energy efficiency. Rating tools allow us to assess houses and set benchmarks and minimum performance levels. They make some green building credentials measurable, which in turn allows them to be regulated by government. Some rating tools are able to predict performance at the design stage. Others measure the actual performance of an existing house with people living in it.

If you are buying an existing house, ask if a rating assessment has been made of the house. This will become more common in the future, but it's worth asking about now. Rating tools for existing houses measure the performance of the house based on its design features and data from energy and water bills. They can give you a ballpark idea of how efficiently a house will run and allow you to compare it with other houses.

# Building a new house

Building a new house is an opportunity to re-draw the size of your ecological footprint. Seemingly small considerations can make a big difference, so make sure you're well informed before making decisions and settling on a site and house plan.

Remember that the fundamental thing to get right is the design of the house in relation to the block of land (often called a 'section' in New Zealand) it will be built on. Good green houses are the product of a perfect match between the design and the block. Many people fall in love with a particular house plan and then plonk it on any block that's the right size. Land can be hard to find, so it makes sense to choose the land first and then design the house to suit the land. There is a common misconception that the only blocks with good solar access are those that have a south frontage, allowing north-facing living areas at the back. In reality, a good architect can design a house with good solar access and north-facing living areas on just about any block. Some examples of good design appropriate to the block of land are included on the diagram on page 30.

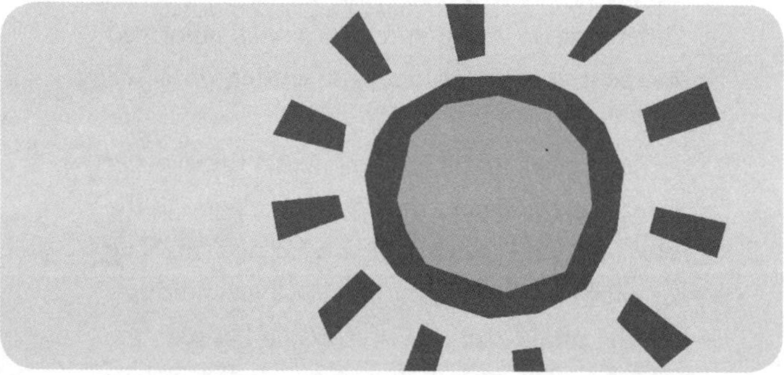

## Environmental ratings for new houses

Energy ratings for houses are particularly important for new homes, the Australian Building Codes Board having announced the adoption of five-star energy efficiency measures for new homes from May 2006. Some states had already introduced minimum energy efficiency rating standards prior to this.

Don't be surprised if you hear mixed opinions on how good or bad five-star standards are. Organisations opposing these improved environmental standards argue that the additional features required to meet the standard make building costs higher for the consumer. However, the increased building costs are offset by reduced running costs. As the saying goes 'the proof of the pudding is in the eating'. In Victoria, where there is a longer history of five-star homes, a survey of the experiences of 150 people living in five-star homes over a 12-month seasonal cycle found that nearly 90% would definitely recommend a five-star energy-rated home to others. Almost the same number, 89%, agreed that all new homes should be five-star rated, 89% said their homes were definitely warmer in winter and 86% said that they were cooler in summer.

When you're consulting a developer or architect, find out the star rating for each design you are considering. Ask about various ways in which the rating can be improved. Don't just aim for five-stars; see how high you can go. There are a number of different tools used by professionals to determine the star rating. Many focus on ratings that reflect the typical energy consumption of the house. However, some look instead at the greenhouse impact of the dwelling, which may be subtly different. Work is being done to diversify these tools so that they can also incorporate other environmental issues and benefits, such as improved water efficiency or indoor air quality differences.

The table following shows some of the main approved rating tools and schemes. You do not need to know these in detail. However, someone trying to sign you up for a new home deal may use terms like these, so it might help to be familiar with the language.

| RATING SCHEME/ TOOL | AREA OF RELEVANCE | ISSUES COVERED | DESCRIPTION |
|---|---|---|---|
| *Nationwide House Energy Rating Scheme (NatHERS)* | Australia | Energy efficiency | An envelope-rating tool that predicts how much heating and cooling energy would be needed to keep the house comfortable. Awards one–five star rating of energy performance |
| *AccuRate* | Australia | Energy efficiency | An updated, more accurate version of NatHERS, launched in 2006. Recognising progress in sustainable design, ratings awarded are up to 10 stars |
| *Building Energy Rating Scheme (BERS)* | Australia, most commonly used in Queensland | Energy efficiency | A similar tool to NatHERS software, based on the same calculation engine. Can incorporate the effects of different ventilation options. Considered a better tool for hot, humid climates, hence popular in Queensland |
| *FirstRate* | Victoria and WA, but can be used elsewhere | Energy efficiency | Developed by Sustainability Victoria, which runs a professional energy rater accreditation program and directory for assessors working in Victoria. Provides an overall rating of zero–six stars. An easy-to-use QuickRate version is also available |

| RATING SCHEME/ TOOL | AREA OF RELEVANCE | ISSUES COVERED | DESCRIPTION |
|---|---|---|---|
| *Energy Smart Home Rating* | NSW | Greenhouse gas emissions | An easy-to-use tool that calculates greenhouse gas emissions of an existing operating building, rather than predicting the energy consumption of a house at design stage. Compares the tested house against the performance of the average household of the same type in NSW |
| *Building Sustainability Index (BASIX)* | NSW | Greenhouse gas emissions and water use | A tool that ensures that new dwellings have reduced greenhouse impact and water use as required by new planning regulations |
| *National Australian Built Environment Rating System (NABERS)* | Australia | Energy, water, wastewater, indoor air quality, waste, material use and more | A performance-based rating system for both homes and commercial buildings. Developed by the federal Department of Environment and Heritage, but run by the NSW government. Covers a wide range of environmental issues, using a whole-of-lifecycle approach |

## Location

Choosing where you're going to live will immediately affect your ongoing greenhouse gas contribution. This has nothing to do with the house itself but with how far the house is from everywhere else that you need to go. The aim is to live locally as much as possible.

Transport accounts for 34% of the greenhouse impact of the average Australian household. The bulk of these greenhouse gases come from the exhaust pipes of cars, so look for a site near a train station or other public transport route so that you don't have to drive. If you live close enough, you could even walk or ride a bike to work in good weather.

Also look at the services available in the area. An established community with shops, schools and other facilities and services will reduce the amount of time you need to spend in your car.

## Planning and design

Before you think about the specific design, think about what you need and want in your house. Consider the number of bedrooms you'll need now and in the future. Factor in storage space and, if applicable, a study or home office. Don't forget to plan for outdoor needs as well: allow space for things like a clothesline, compost bins, a rainwater tank, an outdoor entertaining area, a barbecue and garbage bins, with a path to the front nature strip. Now is also the ideal time to consider landscaping features that might help to improve the energy efficiency of your house, as outlined in *The Green Garden* book in this series.

Remember that you have to pay for every square metre of the building of your home. Each additional square metre will take more materials and energy to build. Once finished, each extra square metre is another to keep clean and another to heat. The frightening trend of building huge McMansions and then having to heat them is part of the reason that Australia's household energy use is still increasing, despite our raised environmental awareness and our efforts to improve energy efficiency. Try to avoid designs that need a lot of hallways for access. Hallways are dead space. Instead, have areas of open-plan living, adequately closed off from the rest of the house to prevent draughts and heat loss. Maximise the amount of useful space, rather than wasting money and resources on hallways and unused rooms. Do your really need a separate home theatre room when there's already a TV, VCR and DVD player in the living room?

## Orientation

Once you've chosen your land, it's time to decide how your dream home will sit on the block and which way the rooms will face. Harnessing the light and energy of the sun will save you a lot of money. The north face of houses in Australia and New Zealand (and the southern hemisphere in general) gets direct sunlight through most of the day. This sunlight can be used to help heat and light your home. Ideally, your block and the placement of the house on it should allow clear access to the north (often called 'solar access'). It should not be overshadowed in winter by other buildings, tall trees or fences to the north. Solar access means your house will get the maximum free light and heat possible in the winter months. It can be restricted in summer by shade sails or deciduous plants.

The benefits of north-facing rooms and windows in New Zealand and the cooler parts of Australia are now widely known. Good use of a northerly aspect is free and will make your home brighter and more comfortable. It will also increase the eventual resale value of your house.

## Home sweet home

In Australia there are over 7.7 million homes. Of these, 80% are separate houses, while around 10% are semi-detached houses, terrace houses or townhouses, and the remaining 10% are flats, units, apartments and other dwellings.

## What goes where?

The location of each room also makes a difference. Consider the following tips.

- Locate rooms in which you spend a lot of time during the day to the northern side of the house. This includes living and family rooms.
- Put bedrooms to the south of the house. This side tends to be the coolest.
- The west of the house gets hot in the afternoon. Put service rooms such as the laundry, bathroom and garage and storage rooms on this side.
- 'Zone' rooms with similar uses and heating or cooling needs close together to make the distribution of heating and cooling easier to control and more energy efficient. In a single-storey house, the three main zones are the living zone, the sleeping zone and the wet areas (kitchen, bathroom, laundry and toilet).

- Place your wet areas as close together as possible to reduce the length of pipes needed. This will also reduce energy costs by shortening the distance that hot water has to travel from the hot water system, losing heat along the way.

**Place wet areas, together with the hot water system, near the kitchen**

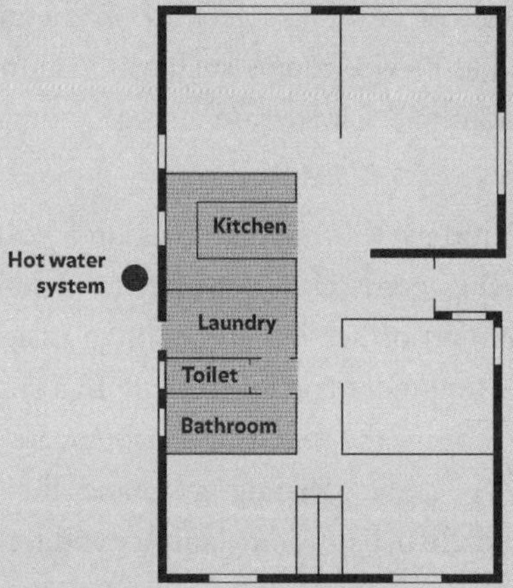

- If need be, place the bedrooms away from the front of the house to keep these sleeping areas away from traffic noise and pollution and to provide privacy.

## Ecological footprint

A person's ecological footprint is a measure of the impact that they have on the environment. It is an estimate of the area of land needed to produce the food and resources they use, and absorb the wastes they produce. The size of your house will have a big impact on the size of your ecological footprint.

## Building materials

The choice of building materials is complicated and best done in consultation with your architect or builder. It requires expert knowledge and will depend upon the design of the house, the local climate and the condition of the land you're building on. From an environmental point of view the main considerations are:

***Material properties*** Different building materials have different properties that make them suited to particular climates and uses. For example, a humid climate will need materials that are moisture-resistant while coastal areas will need materials that resist corrosion.

***Durability*** As a general rule, the longer a building lasts, the less impact it has on the environment. Materials chosen should be durable, or easily replaceable should they have a shorter life span than the building as a whole. Often, higher quality materials will cost more than cheaper alternatives, but they will save you money in the long term by needing less maintenance, not needing premature replacement and improving the house's resale value.

***Renewable or recyclable resources*** Plant-based materials are renewable in that they can, in theory, be replaced by growing more, but our supply of resources such as metals is limited. It's important to make sure that, whenever possible, non-renewable materials can be easily recycled at the end of the building's life span.

***Forest-friendly timbers*** Some natural building materials, particularly forest timber, are harvested from areas of natural vegetation. This destroys habitat, harms wildlife and has an impact on biodiversity. Always question the source of any timber that is used in your home and ask for alternatives, such as recycled or plantation timber. For further information on choosing the good wood, see page 40.

## Put living areas on the northern side of your house

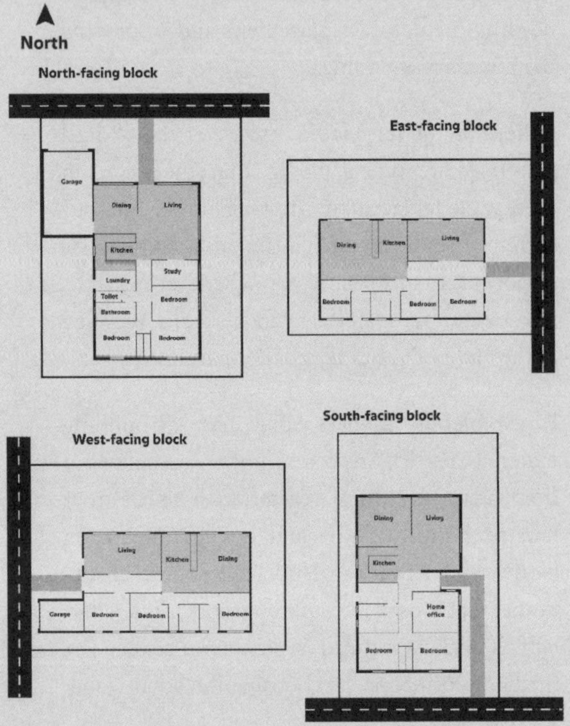

**_Energy efficiency_** Materials have varying degrees of energy efficiency too. For example, materials that have a high thermal mass (concrete, mud brick, rammed earth, stone and slate) are slow to heat and take a while to cool down. They help to regulate the temperature of your home by capturing and storing the sun's heat during the day and releasing it at night. High thermal mass materials can be used for floors in living areas that catch direct winter sunlight. The radiant energy from the sun comes through the windows during the day and is stored in the thermal mass of the floor. It is gradually released during the night, providing passive solar heating. In climates with cool winters, heavy building materials inside with high thermal mass help to capture and store the sun's daytime warmth. For this reason, good architects will consider the size and placement of windows in conjunction with the use (or otherwise) of high thermal mass floor materials.

***Indoor air quality*** Synthetic building materials, finishes and paints can all give off high levels of polluting gases, including volatile organic compounds (VOCs). These pollutants pose a significant health risk and can bring about symptoms such as headache, fatigue, respiratory problems, dizziness and eye, nose, throat and skin irritation. Where possible, limit your use of carpets (particularly those made from synthetic fibres), synthetic adhesives, paints, varnishes (particularly those containing polyurethane) and other finishes as well as particle board and MDF board. It can be very difficult to build a new house totally free from materials that give off these gases so make sure that you ventilate the house regularly and thoroughly for the first year, particularly the first six months.

***Waste and recycling*** Some materials and building methods generate a high amount of waste, while others can actually use recycled waste. Australians are among the highest producers of waste per capita in the world. Up to 40% of the waste going to Australian landfills comes from the building sector. Wherever possible, choose building or landscaping materials that are made from recycled aggregate or other recycled materials. Ask your builder about their waste minimisation practices and consider incorporating waste minimisation and recycling clauses and incentives into the contract. You can also use building materials and features that have been salvaged from demolished buildings. Reclaimed bricks, for example, can add a lot of character to a home or garden and are sometimes cheaper than new materials. Keep an eye out for unusual or interesting pieces that can bring interest and individuality to your home.

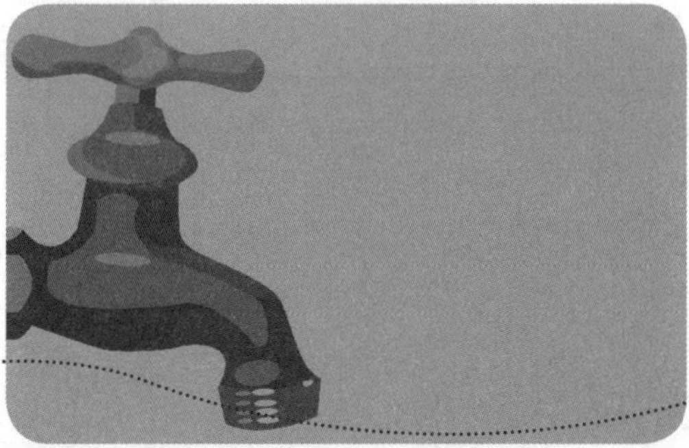

## The house envelope

The barrier between the inside of your house and the outdoors is the house envelope. Windows, doors, walls, roofs and floors make up this envelope, protecting you from the elements, temperature extremes and some of the forces of nature. The ideal envelope is well insulated and airtight, has windows chosen to balance the need for natural light with the need to control heat loss, and allows controllable ventilation to ensure healthy indoor air.

***Insulation*** Insulation is essential to keep your home at a comfortable temperature throughout all seasons. Depending on the design of the house, it can be used to line the roof, ceiling, attic, external walls, basement, crawl space and floor. You may also wish to consider acoustic insulation to block out noise pollution, particularly if you have a music room in the house (perhaps with a drum kit) or you live near a busy road or railway line.

# Heat-leaking houses

Whenever you turn on an airconditioner or heater in a poorly insulated home, your money disappears into thin air.

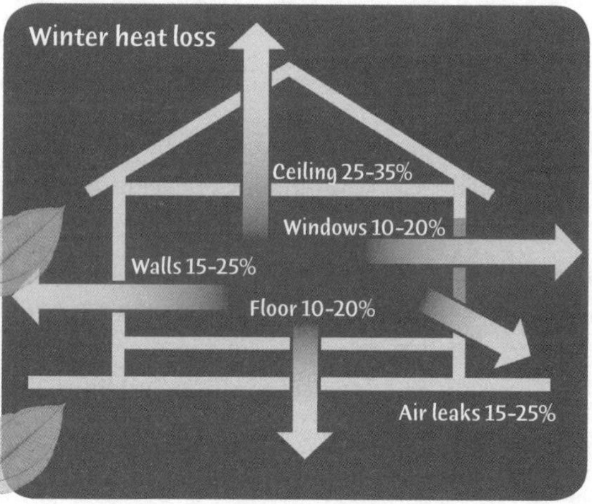

Soft furnishings, such as rugs, curtains and wall hangings, can also provide some insulation.

Insulation comes in two types, bulk and reflective. Bulk insulation uses porous materials that trap air in small pockets to reduce the flow of heat. Just about any material can be used to make insulation, as long as it can trap air. Many insulation products are made from recycled materials, making them even greener. Reflective insulation, as the name suggests, works by reflecting some of the radiation that falls on it. Both types can be used in combination to maximise the insulating effect.

Some people try to skimp on insulation, especially when the building budget is tight. This is a false saving as you pay more in heating and cooling bills over the time you live in the house, not to mention the suffering from being too hot or too cold. Particularly consider insulation for external walls and floors when you're building because it is harder to install in existing constructions than ceiling and roof insulation.

Insulation is made and sold in imperial measurement (R) values, a measure of the insulation's resistance to heat flow. The higher the R value the better insulation the material gives. The Australian Greenhouse Office recommends certain levels of insulation by R value for walls, basement walls, ceilings, solid roof decks and floors in different climate zones in Australia. They can be found online at www.greenhouse.gov.au/yourhome in the 'Insulating for comfort' section of the Consumer Guide. The table on page 39 shows recommended levels of insulation for homes in some of Australia's major towns and cities and various regions of New Zealand.

If your house is built on a concrete slab, the edge of the slab should also be insulated, as a single square metre of exposed slab can lose as much heat as several square metres of uninsulated wall. This is particularly important in homes with in-slab heating systems, which can lose even more heat and waste huge amounts of energy.

Insulation should be considered as an integral part of the whole house envelope. Considering the amount of heat loss from large, single-pane windows, it would be pointless to add a lot of extra wall insulation to a house that has this type of windows.

***Windows and shading*** Windows are a mixed blessing. On the one hand they allow light and warmth in during winter. They also let fresh air in and pollutants out. On the other hand, they permit undesirable heat transfer. In fact, windows can be the cause of around 10–20% of heat loss in winter and around 25–35% of heat gain in summer.

Windows can be made more energy efficient through the addition of an extra pane of glass, known as double glazing, which stops some of the heat transfer. They can also be further insulated with thick curtains and well-fitting pelmets.

Recent years have seen technological developments with windows that further improve their

## Recommended insulation levels (R values)

| | ROOF/ CEILING | WALL |
|---|---|---|
| *Melbourne, Vic.* | 3.0 | 1.5 |
| *Canberra, ACT* | 3.5 | 1.5–2.0 |
| *Hobart, Tas.* | 3.5 | 1.5–2.0 |
| *Thredbo, NSW (alpine climate example)* | 4.0 | 1.5–2.0 |
| *Darwin, NT* | 0*–4.0 | 0*–2.0 |
| *Brisbane, Qld* | 1.5–2.5 | 1.0 |
| *Perth, WA* | 1.5–3.0 | 1.5 |
| *Alice Springs, NT* | 1.5–4.0 | 1.5–2.0 |
| *Sydney, NSW* | 1.5–3.0 | 1.5 |
| *Adelaide, SA* | 2.0–3.0 | 1.5 |
| *North Island excluding the Central Plateau, NZ* | 1.9–3.3 | 1.5–2.6 |
| *South Island, including the North Island Central Plateau, NZ* | 2.5–3.5 | 1.9–2.6 |

* The zero figure doesn't mean that no insulation is required. It means that insulation to reduce heat loss may not be cost-effective and can sometimes contribute to overheating in homes without adequate sun control.

# Timber: choosing the good wood

You can help to preserve our planet's precious rainforests and old-growth forests by refusing to use the timbers that are harvested from them. Instead, look for alternative and sustainable timber choices. Here's a quick guide to recommended timber choices.

### Plantation timbers

Commonly available plantation species include:

- hoop pine (*Callitris collumelaris*)
- slash pine (*Pinus elliotii*)
- radiata pine (*Pinus radiata*)
- blue gum (*Eucalyptus globulus*).

## Recycled timbers

- Commonly available species include blackbutt, alpine ash, jarrah, messmate, mountain ash, cedar, rosewood, silky oak, Victorian ash and many others. Visit www.timbershop.org to search online for recycled timber stocks.
- Note that recycled timber does *not* include timber salvaged from land-clearing or forestry operations or dead native trees that might provide wildlife habitat.
- Avoid recycled timber that is chemically treated or impregnated with pesticides.

## Agro-forestry timbers

Agro-forestry timbers are those established and managed as part of an agricultural enterprise, such as:

- blue gum
- cypress
- radiata pine.

**Timber alternatives**

- bamboo—commonly used in flooring and furniture
- recycled plastic 'lumber'—often used in decking
- wheat-based fibreboards

There are also a small number of timber suppliers that are certified in line with the Forest Stewardship Council (FSC) standards. FSC is an international non-profit organisation working towards the environmental, social and economic management of the world's forests. FSC in Australia is in its early days, but hopefully we will see it become prominent and established Down Under. For more information visit www.fscaustralia.org or the 'Forests' section of the WWF site at wwf.org.au.

Be wary of other timber eco-labels as many are self-evaluating voluntary initiatives run by the forestry industry and may not require independent auditing.

performance over the old, cold conventional single panes. One development is the low-emissivity (low-e) coating—a thin transparent layer of metal oxide over the glass that allows visible light through but reflects back infrared heat radiation. This helps the window to keep heat out in summer and reflect heat back into the house in winter. It also has the added benefit of reducing incoming UV light and so protects furniture and drapes from fading. The space between panes in multiple-glazing can also be filled with argon gas, as inert gas that provides more insulation than air. Gas filling in combination with a low-e coating is particularly effective at improving the insulating level of windows.

Size and placement of windows and how they are shaded will depend on their orientation and the room they're placed in. The materials the window and its frame are made of, and how well they are fitted, will also influence how well the windows perform.

## Tips for windows

- If there is good solar access, use large windows for north-facing living areas.
- Keep windows on the south face of the house a small to medium size.
- Put small windows on the east and west faces of the house, with good summer shading. The east gets direct sunlight in the morning hours, while the west gets hot, direct sun in the afternoon.
- When selecting windows, look for the Window Energy Rating Scheme label, which is like the energy rating labels for whitegoods. For window recommendations for your climate area (within Australia), go to the scheme's website at www.wers.net. Though the program doesn't strictly cover New Zealand, there is useful information for New Zealanders at this website in the section for Heating Climates (climate areas that require more heating than cooling).

- Position windows that can open and close in places that allow for cross-ventilation in living areas.
- Use double glazing for windows that are high up and hard to get at, such as skylights and clerestory windows. These windows may be hard to put curtains on or may not be intended to have curtains for aesthetic reasons.
- For the framing material, metal frames (usually aluminium) without a thermal break allow heat loss and gain. If you choose metal frames, make sure that they have some form of insulating treatment. Wooden frames insulate well but need more maintenance than aluminium. PVC frames are low maintenance and insulate well, but are made from non-renewable resources and contribute to pollution in their production. PVC in general should be avoided.
- Always seal gaps between the window frame and the wall to reduce air leaks, draughts and heat loss.

- Fit windows that can be opened with flywire screens to keep insects out and to remove the temptation to use bug spray.
- Use window coverings that will make your house more efficient. Thick insulating curtains with pelmets will keep warmth inside in winter and block out the summer sun.

## Tips for shading

- Make sure that your roof has eaves that overhang the north face of the house. During winter, the path of the sun is low enough in the sky to allow light and radiant heat into the room. However, in the summer, when the sun is higher, the eaves will block the sun and help to shade the room.
- External blinds can effectively prevent much of the summer heat coming through the windows. Particularly consider them if you don't have much shade on the west side of the house.

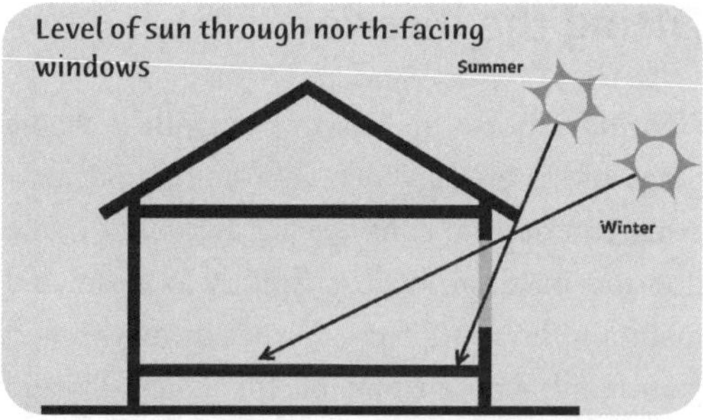


- Deciduous plants can provide shade in the summer and lose their leaves and allow the light through in the winter. Consider planting these to the north of your house or near windows on the east and west faces of the house. Removable shade sails or shadecloths can similarly provide shade in the summer and be taken down in the winter.
- If you live in a hot climate, where northerly solar access is undesirable, consider building your house with a veranda encircling it.
- You can also use a DIY window-tinting kit, available from some hardware stores.

## Fitting the interior

***Flooring*** Floors can be covered with a whole range of materials. When choosing a floor covering you need to consider the durability of the flooring material, how easy it is to clean and maintain, how it affects the health of you and your family, how sustainable the material is and the special needs of particular rooms for floor ing. Some materials are better suited to some rooms than others.

**Floorboards** are easier to clean than carpet and discourage dust mites. They can also be better for the quality of indoor air compared with gas-emitting or dust mite-ridden carpet, provided they're not finished with an off-gassing synthetic varnish. If you're looking at timber floorboards, consider the option of using recycled or plantation-grown timber.

**Carpets** are often preferred in bedrooms because they're softer on bare feet, absorb noise and help to keep out draughts. However, they can provide the right conditions for dust mites to grow in, so carpets shouldn't be used if you have family members with respiratory problems. Synthetic carpets, like other synthetic materials, produce gas pollutants, and even pure wool carpet often has a synthetic underlay or chemical treatment that can contribute to poor indoor air quality. Consider washable rugs as an alternative.

**Natural fibres**, including sisal, coir and jute, make sustainable and low-allergy floor coverings.

**Resilient flooring** is soft, rubbery-feeling flooring that comes as tiles or rolls of sheet flooring. It is generally made from vinyl, rubber, cork or linoleum (lino). Resilient flooring comes in a wide range of designs, finishes and prices. It is often used because it is fairly easy to install and maintain and is relatively cheap to replace. However, it can be thin, easily damaged and can show irregularities in the surface over which it is laid. Care must be taken when choosing the adhesives that stick the flooring to the subfloor as some adhesives off-gas high levels of VOCs.

# Eco-carpet

If you're looking for carpet for a school, government or commercial building, have a look at the award-winning Interface Flooring systems. Interface has a unique and sustainable approach to providing flooring systems, providing carpets made from recycled materials, suited to high-traffic areas and able to replace and recycle just the worn-out carpet tiles (leaving carpet tiles in good condition in low-traffic areas in place).

Visit www.interfaceaus.com.au for more information.

Look for water-based adhesives where possible. Resilient flooring products made from the natural materials cork, rubber and lino are renewable and sustainable. Vinyl and other PVC products are generally polluting in their manufacture.

**Cork** is a kind of resilient flooring with a wonderful environmental story behind it. Cork is a natural, sustainable material harvested every nine years from the bark of the cork oak tree. It grows without the need for chemicals, fertilisers or irrigation. Cork is also non-polluting, biodegradable and recyclable. What more could you ask for? Cork tiles help to insulate the floor and are soft to walk on. If you decide to use cork flooring, avoid tiles that are PVC-finished. Consider buying them unvarnished and finish them yourself with a water-based varnish or beeswax polish. The fact that they come in tiles makes it easy to replace any sections that become damaged, so make sure you buy a few extra.

**Bamboo** is another sustainable natural floor covering. It is an incredibly fast-growing plant, making it a renewable and quick-to-produce green alternative to timber floorboards. Bamboo floorboards function similarly to hardwood floorboards. Their key benefits are that they are moisture-resistant, very strong, durable, don't buckle or twist, are free of knots or large flaws, and are naturally a light blond colour that can be stained to produce the effect of darker woods if desired. Bamboo flooring is cost-effective and readily available in tongue and groove strips. It can be bought both unfinished and polyurethane-finished.

**Slate, stone, ceramic and terracotta tiles and exposed concrete** are popular flooring materials, particularly in the wet areas of the house. Because they are made from mineral resources mined from the earth's crust, they are not renewable, but some of the raw ingredients, like the sand in concrete, are in abundant supply. Others, such as some kinds of slate, granite and marble, are more rare. The main environmental impacts from these materials are the damage done to local ecosystems when they're extracted,

## EcoSpecifier

Want to find some greener building and interior materials and don't know where to start looking? The RMIT University Centre for Design has produced a wonderful resource called the EcoSpecifier. This is a fully searchable online database of environmentally preferred materials and where to get them. It includes listings for recycled plastic and hemp textiles, bamboo flooring and a huge range of other housing materials. Each listing gives an overview of the environmental benefits and where the materials can be used. Go to www.ecospecifier.org.

## Green homes on display

GreenSmart is a joint initiative of Housing Industry Association Ltd (HIA) and the Australian Greenhouse Office that encourages home buyers and builders to build houses that are more environmentally responsible. There are a number of HIA GreenSmart Display Villages around Australia, where prospective home buyers can see display homes that incorporate sustainability. For example, Parkville Gardens is an inner city village of 6-star-rated homes, which was completed in time to serve as the Melbourne 2006 Commonwealth Games Athletes' Village.
More information:
www.greensmart.com.au

the energy and water used and waste produced in their processing and, being heavy materials, the fuel and greenhouse costs of their transportation. Materials that have to be fired in a kiln, such as ceramic tiles, also have a high energy cost.

For these reasons, stone and ceramic flooring products are a good choice environmentally when they're recycled or salvaged stocks, sourced from a local supply. Other benefits are that they are generally easy to clean without the need for harsh cleaning agents and they also provide better indoor air quality than carpets, as long as they haven't been finished with fume-emitting varnishes or sealants. They don't harbour dust mites.

You can also get various kinds of aggregate or engineered stone, a mixture of mineral materials (such as quartz, marble or granite) and binding resins. They tend to be long-wearing and are often made in non-porous varieties. One environmental benefit of engineered stone is that it can be made with some recycled or salvaged content, such as waste marble, recycled coloured glass and even oyster and mussel shells.

***Green fittings*** Nowadays there's a lot more to a house than four walls and a roof. Most houses also come with heating and cooling systems, a hot water system, cooking appliances, toilets, baths and showers and a range of tap fittings. All of these can be chosen to reduce your ongoing use of energy and water. Here's a snapshot of what to consider.

## General rules for heating systems

Incorporating heating needs into a building design is the ideal approach to heating. There is a huge range of different heaters and heating systems, with individual models having their own features and efficiencies. Details of the options available are explored in the book *The Australian Green Consumer Guide*. The size and shape of a house, or even just a single room, will determine the type of heater that will most efficiently heat it. However, a few generalisations can be made.

- Convective (air) heaters heat the air and circulate it. They are better for heating small rooms quickly. Because convective heaters warm the air itself, rooms with a lot of air (in other words, large rooms) or with open stairwells and other openings are not suited to convective heaters.
- Remember also that hot air rises, so a convective heater in a room with a high ceiling will not provide heat efficiently for people at ground level, although a ceiling fan can help to recirculate the warm air that collects at ceiling level.
- Radiant heaters heat objects rather than air. Examples include in-slab heating, hydronic heating and bar radiators. They can take a while to warm up, but are better for larger spaces and draughty rooms.
- Particularly consider whether or not you would like an in-floor heating system, such as hydronic or electric slab heating, or a

ground-source heat pump. These generally cannot be fitted after the house is built without considerable cost and work.

- If choosing a central heating system, see if you can have sleeping and living areas or (in the case of multiple-storey houses) different levels separately zoned so that you can control the heating in the areas individually and switch off zones that don't need heating.
- Choose systems that have a remote, programmable electronic thermostat, allowing you greater control over heating.
- Pilot lights in gas heating systems can cost up to $25 per year extra. Choose systems with electrical ignition instead, or at least turn the pilot light off in summer when the heater is not in use.
- Gas heating and efficient reverse-cycle airconditioners (also called heat pumps) are more efficient, accounting for a third of the amount

of greenhouse emissions of other electric heating systems.

- Hydronic heating systems, which circulate heated water to provide heat for rooms, can be connected to solar hot water systems (preferably gas boosted) to source some of the heat from the sun.
- Some heating appliances bear either the Energy Rating label (which includes reverse cycle air-conditioners) or the Gas Energy Rating label. Appliances are rated for energy efficiency on a scale of one to six stars. The more stars a product has, the more efficient it is, the less electricity it needs to do its job and the lower its ongoing running costs. Use the rating labels to compare products. For more information visit www.energyrating.gov.au.

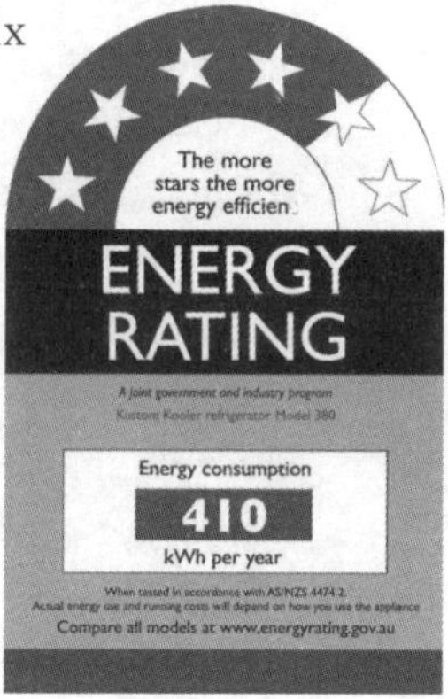

## Cooling systems

- Good design will prevent your house from capturing too much heat and allow it to catch cooling breezes. Together these things will reduce your need for active cooling. Depending on the climate of where you live, you may be able to get away with natural ventilation or a simple fan to cover your cooling needs. Factor cross ventilation into your home design plans.
- Fans, evaporative cooling and refrigerated air-conditioning are the three types of active cooling. All run on electricity and therefore have an energy cost and a greenhouse impact.
- Fans use the least amount of electricity, followed by evaporative coolers.
- Evaporative coolers are best suited to dry climates.
- Airconditioners cool the air and remove moisture, but need large amounts of electricity to run. They are covered by the Energy Rating label program, so use the ratings to compare systems.

## Hot water systems

- In many parts of Australia, gas-boosted solar is the greenest choice of hot water system.
- Locate the hot water system as close as you can to the kitchen to reduce the amount of cold water sitting in lengths of pipe that gets wasted while waiting for the hot water to come through. Hot water will also lose more heat travelling through a greater length of pipe on its way to the place of use.
- If you're considering a solar hot water system, make sure you factor this into the roof and plumbing design for the house.

## Water-using devices

There is also the Water Rating label for water-using devices. Recognising the need to conserve national water supplies, the Australian government, in collaboration with state and territory governments, has introduced a Water Efficiency Labelling and Standards (WELS) Scheme, which applies national mandatory water-efficiency labelling (and a minimum performance standard to toilets) to household water-using products. Like the Energy Rating label, the Water Rating label rates products on a scale of one to six stars, with a greater number of stars indicating a more water-efficient product and a better environmental choice. Once again, the more stars the better.

The Water Rating label also states a water consumption figure. This provides an estimate of the amount of water the product typically uses when tested under Australian Standards criteria. For example, the Water Rating label for a showerhead will state water consumption

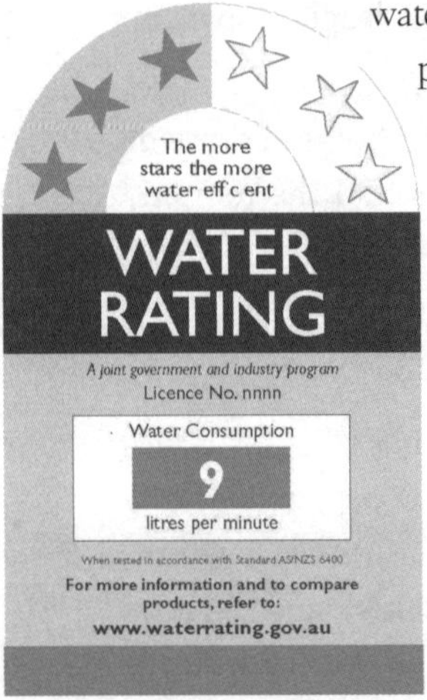

as a flow rate in litres per minute, while a washing machine will state water consumption in litres per wash. The Water Rating label can help you to choose toilets, tap fixtures, other plumbing devices and water-using whitegoods that will help you cut your water consumption and, in doing so, save a little money on your water bills.

The Water Rating label takes on even more importance with devices that use hot water. High-rated, water-efficient products reduce the use of water and the energy needed to heat them. As a result, water ratings can also help you to reduce your energy costs. For more information, visit www.waterrating.gov.au. You can also use the Water Rating search engine to compare the water ratings of available products and calculate estimates of a product's lifetime water consumption. Visit http://search.waterrating.com.au.

## How hard is it to build green?

Fortunately, the government now wants us to build more energy-efficient houses so that we can make an effort towards reducing our greenhouse emissions and attempt to meet the Kyoto Protocol targets. Following Victoria's leading five-star rating requirement for new dwellings, five-star efficiency will soon become the norm for home builders. More builders and developers are seeing the competitive edge that greener, more comfortable, more energy-efficient homes have, and are developing project homes with higher energy ratings.

Look out for display homes and house plans that are designed with energy or the environ-

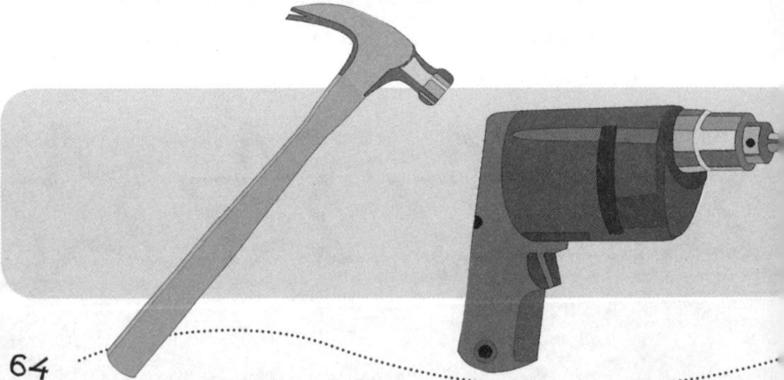

ment in mind. Don't assume that the designs on offer are set in concrete and can't be changed. You'll find that developers are often willing to modify standard plans, for instance by changing the placement of windows or increasing the size of eaves at little or no additional cost. You can also pay a little extra to incorporate other environmental features, such as additional insulation or upgrading the hot water service to a higher efficiency or solar system. Don't be afraid to ask. Even if they say no, your interest in environmental concerns will often be registered and they may consider designing greener alternatives in the future. Change often begins with enlightened people asking for it.

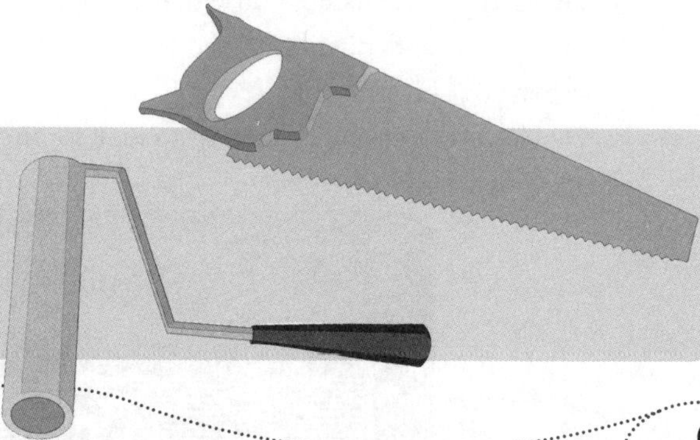

# Buying an existing home

When you're looking at existing houses to buy, remember that renovations can turn a moderate house into a greener house, but all renovations come at a cost, in both time and money. Don't forget that it's the existing house that you'll take possession of and possibly move into, not the house in your imagination.

Here's a green checklist of what to look for in an existing house.

## Green checklist for house hunting

### Energy

- How much natural light is there in the living areas?
- Is the house draughty? Are there cracks or gaps in the walls or floors?
- What is the current heating system? How efficient is it?
- What other energy-using systems are in place?
- If it's an older home, have energy-efficient improvements or renovations been made?
- Overall, how energy-efficient is the house and how much electricity and gas does it typically use? You can ask to see past energy bills to get an indication of what it costs to run, taking into account the previous number of occupants and their particular living habits and appliances.
- Is the house well zoned to make heating and cooling easier?
- How well is the house insulated? Can more insulation be added if needed?
- Are the windows single pane or double-glazed?

## Design

- Does the design suit the climate? For example, if it's a cooler area, is there good solar access, or if it's in an area that gets hot summers, is the house adequately shaded with eaves or verandas?
- Are the living areas on the north side?
- Are the bedrooms on the south side?
- Are the service rooms on the west or south sides?
- Does the placement of windows allow for quick cross-ventilation?

## Interior

- What are the floors covered with? Are the coverings in good condition?
- What appliances are included in the goods and chattels?
- Have any toxic materials been used inside, such as lead-containing paint?

If you come up with satisfactory answers to most or all of these questions for a house you're considering, buy it. If not, then weigh up what is lacking against your ability and budget to change it. Sometimes it can be better to buy a smaller house that's on the right track and in the right location and add an extension later. Always get professional advice and an estimate of the costs of changes by having a professional builder or architect look at the house before you buy.

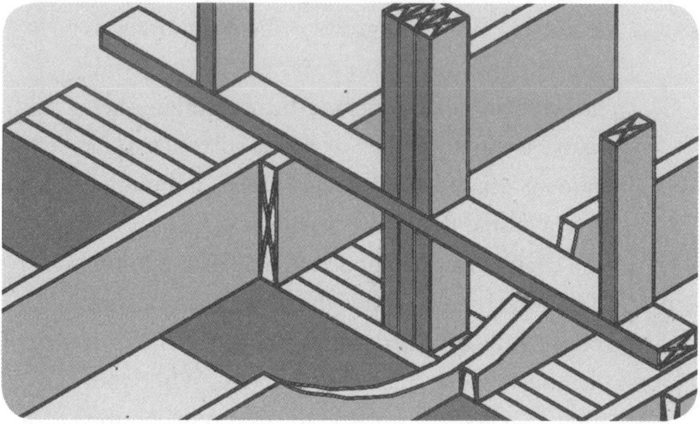

# Green renovations

Renovation is a way to bring new life to an old home. It's also a way to stay in a loved home as a growing family outgrows it. Green renovations can improve the comfort and efficiency of your home. This might include incorporating alternative technology, such as a new solar hot water system. Many of the green ideas for building a new home that we've seen in the first part of the chapter can be applied to an existing house. The added dimension is that there is an existing house with solid structures to change, move or demolish, rather than a blank canvas. While your aims may include helping the environment, this process can instead harm it. A few simple green renovating tips will help you to change your house into your dream home without producing an unhealthy home environment or needlessly harming the planet.

## Bringing in an expert

For most people, their house is the biggest purchase they will make in their lives and is their greatest asset. As a home owner, it's up to you to make sure that your house is safe, in good condition, healthy to live in and not too expensive to run. Your house may contain an accident waiting to happen or a heater guzzling more energy than need be—all without you knowing it. Just as a doctor or naturopath can give your body a check-up and uncover any undiagnosed health problems, a professional home inspector, builder or architect can give the house you live in or one you're considering buying a building check-up. It is your responsibility to know or find out what state your house is in.

In particular, Archicentre architects can provide professional, independent services to home buyers, home builders and home renovators. Archicentre is the building advisory service of the Royal Australian Institute of Architects.

Archicentre offers pre-purchase home inspections, in which an architect carefully inspects the house using a comprehensive 300-point checklist. The report identifies any components that aren't performing properly, safely or efficiently, any items that need replacing or repairs and any health and safety risks. You can also use the report to help in negotiating a purchase price for the house.

Archicentre pre-purchase house inspections generally cost around $500 for the average three-bedroom home (more or less, depending on the size of the house) and take around an hour. They need to be booked around two working days in advance. In some areas, Archicentre also offers home safety inspections. Visit www.archicentre.com.au for more information.

## Green renovating tips

***Sanding*** should be done with a sander with a vacuum attachment. Wear a mask to protect your lungs. Take particular care if you're sanding wood that has been treated, as the fine wood dust particles will also contain potentially harmful chemicals.

***Paint stripping*** should be done with water-based, low-allergy paint strippers. Avoid conventional paint strippers, particularly those that contain dichloromethane (DCM), which is believed to be carcinogenic.

***Paint containing lead*** may be found in houses built before 1970. Lead is a highly toxic element. Exposure to even the tiniest amounts of lead in dust or debris from renovations can cause health problems. Hardware stores sell home kits that help to test for the presence of lead. If your home does have paint containing lead, get professional advice before removing or disturbing it. If you do plan a renovation that will disturb it, then move out of the house until the renovation is finished. The Department of Environment and Heritage has information about lead house paint online at www.deh.gov.au/— look for 'lead' under 'L' in the site index and click on the link 'Lead in House Paint'.

***Wallpapering*** can retain moisture and lead to fungal growth. Mix borax into wallpaper paste to prevent this instead of using a fungicide.

***Painting*** can be made less wasteful by getting your estimates right. Choose plant-based paints and finishes as they're made from sustainable materials rather than petrochemicals. They also produce far fewer indoor air-polluting fumes. Plant-based paints avoid using volatile petrochemical solvents and consequently can take long to dry. Allow for this when planning your renovations.

***Don't paint*** Consider opting for natural, unpainted finishes where possible. Some porous materials such as cork and wood can be adequately sealed and preserved with natural oils or beeswax.

***Recycle*** Carefully remove any slate or ceramic tiles, aiming to keep as many intact as possible. Recycled tiles can often be sold second-hand or reused in landscaping or other areas of the house. If you feel like getting crafty, keep broken tiles for making mosaics.

***Clean up safely*** Never pour unused paint, solvents or water from washing brushes into stormwater drains as this can seriously contaminate ground water and nearby streams, wetlands,

lakes, rivers and oceans. Unwanted paint and solvents should be disposed of in an environmentally responsible manner. Some states, regions and local councils have hazardous and liquid waste collections, so contact your local council as the first port of call for specific advice. Unused paint can be disposed of by brushing left-over paint onto sheets of old newspaper. Brush one layer, then put the next layer directly on top and brush that one too, and so on. Let the paper dry and throw it out with your normal rubbish. In some areas, dry steel paint cans can be collected for recycling with other steel food cans.

Always remember to clean up thoroughly after any renovating, making sure that you vacuum and clean up every skerrick of dust or spilt paint. Over the following days, vacuum frequently and air the house to make sure that you capture any solid particles that have escaped you earlier and to allow any potentially harmful fumes to dissipate. That way you can enjoy your greener, new-look home without risk to your health.

# Epilogue: A Step Further

Next time you go to the beach, go hiking, go camping or take a trip to the zoo, look around you. Our planet is a beautiful, unique place that is well worth saving. The power to do so rests in your hands. Apart from the things you can easily change on a daily basis, there are other things you can do to make a difference.

## Get informed

If you care about the state of the environment it's important to be informed about what's happening around the globe, what the problems are and what's being done about them. Without awareness and information the green movement would not exist.

An easy way to stay up-to-date is to read news extracts from the Planet Ark/Reuters Daily World Environment News Service. Sponsored by Reuters, Planet Ark publishes up to 40 environmental news stories, written by Reuters journalists worldwide, each weekday on the Planet Ark internet site. This site covers a range of green topics, from A (air pollution) to Z (zoo news). You can also subscribe to the service for free and receive a daily email newsletter containing the day's news headlines and links to the full stories. Visit www.planetark.org/news/.

## Get active

Get involved in practical conservation efforts, planting trees and restoring natural vegetation. You can call your local council to find out about local Landcare and other conservation groups. Alternatively, visit Volunteer Green, an initiative of the Australian Conservation Foundation that coordinates a broad range of volunteer efforts. For more information visit www.volunteergreen.org. For opportunities in New Zealand, contact the New Zealand Trust for Conservation Volunteers (www.conservationvolunteers.org.nz)

or the New Zealand Landcare Trust (www.landcare.org.nz).

You can also join a number of community-based environmental monitoring programs, such as Waterwatch Australia (www.waterwatch.org.au), a national community water quality monitoring network that involves ordinary Australians in the protection of their waterways. For animal lovers, there are also wildlife monitoring programs, such as the Birds in Backyards program (www.birdsinbackyards.net), a joint initiative of Birds Australia and the Australian Museum.

If you don't have time to do it on a regular basis then join a one-off conservation effort, such as Planet Ark's National Tree Day (www.planetark.org/treeday) or Clean Up Australia Day (www.cleanup.com.au).

Donations to green groups and conservation funds can also make a difference. However, always remember that what the planet really needs from you is lifestyle change, not loose change.

## Get political

Don't be afraid to speak out about the issues that concern you. Write letters or emails to the relevant politicians and government departments and let them know how you feel. However, don't just protest, offer your thoughts on solutions too.

## Pass the message on

Help your friends and family to green up their lives. Share your favourite eco-tips with them. The more people we can enlist to help the environment from the comfort of their own homes, the bigger difference we can make to the health of the planet.

# Notes

**General sustainable building:** numerous interviews with Dr Graham Treloar, School of Architecture and Building, Deakin University, and Deputy Director of the Built Environment Research Group and Paul Klymenko, Director of Research, Planet Ark. General sustainable building information from the Australian Greenhouse Office publications *Global Warming, Cool It!* and *Your Home: Good Residential Design Guide*; Western Power publication *Smart Home*; HIA GreenSmart initiative; SEAV brochures *Energy Smart House Design* and *Energy Smart Renovations*; *Sustainable House* by Michael Mobbs' (Choice Books, 1998); Alternative Technology Association's *The Green Technology House and Garden Book* (1993). **Victorian experiences of five-star homes:** Media release from Victorian Minister for the Environment, John Thwaites: 'Overwhelming support for 5 star homes', 7 April 2006. **Material resources:**

RMIT Centre for Design (particularly their publication *EcoSpecifier*); The Wilderness Society's One Stop Timbershop. **Housing statistics:** Australian Bureau of Statistics. **Insulation levels:** Insulation level recommendations for Australia from Australian government publication, *Your Home*, 3rd edn, 2005; for New Zealand from the Energy Efficiency and Conservation Authority and Standards New Zealand. **Windows:** Window Energy Rating Scheme; Sustainability Victoria's brochure 'Energy Saving for Windows'. **Floor covering:** *Our House* television program factsheet 'Bamboo Floor'; cork information from Amorim Group and Cork Masters; other material information from the RMIT *EcoSpecifier*; *Green Living* by Sarah Callard and Diane Millis (Carlton Books Ltd, 2001). **Dichloromethane:** listed as a suspected carcinogen by the US EPA.

# Further information

## Environmental design and building resources

### NATIONAL (AUSTRALIA)

**The Australian Greenhouse Office—*Your Home***

Everyone who is considering building, buying or renovating should get a copy of the AGO's *Your Home*. It can be viewed online or ordered as a hard copy or DVD through the Alternative Technology Association.

The Consumer Guide is an easy-to-read introduction to green building for the average person. The Technical Manual has further information and factsheets for builders and contractors.

www.greenhouse.gov.au/yourhome/index.htm

For other information on building energy efficiency go to www.greenhouse.gov.au/buildings.

**National Australian Built Environment Rating System (NABERS)**

NABERS rates a building on the basis of its measured operational impacts on the environment. NABERS is a national initiative managed by the Department of Energy Utilities and Sustainability NSW government.
www.nabers.com.au

**GreenSmart homes**

A joint initiative of the Housing Industry Association Ltd (HIA) and the Australian Greenhouse Office that encourages home buyers and builders to build houses that are more environmentally responsible. The GreenSmart website has a list of accredited builders. GreenSmart is a package of practical environmental measures that minimise building waste and improve water and energy efficiency, both during the construction and use of the home. The program is localised to various geographic regions of Australia, ensuring that information is relevant to its climate zone. There are eight offices across Australia.
www.greensmart.com.au

**Green Building Council of Australia**

A national, not-for-profit organisation, supported by industry and governments across the country. It aims to develop a sustainable property industry for Australia and drive the adoption of green building practices through market-based solutions. Runs the Green Star rating system for commercial buildings, with plans to develop Green Star tools for Retail, Health, Education, Convention, and Residential buildings.
www.gbcaus.org

**Archicentre**

The building advisory service of the Royal Australian Institute of Architects. Archicentre professionals can conduct pre-purchase house inspections for prospective home buyers as well as home safety inspections.

www.archicentre.com.au

Ph: 1300 13 45 13

## NSW

**Energy Smart Home**

Visit the website to conduct a virtual home audit.

www.energysmarthome.com.au

BASIX—the Building Sustainability Index

This NSW government measurement tools and certification program ensures homes are designed to use less potable water and be responsible for fewer greenhouse gas emissions by setting energy and water reduction targets for house and units.

www.basix.nsw.gov.au

## QLD

**Sustainable Homes**

An industry and Queensland government-backed initiative aiming to provide communities throughout the state with display homes that incorporate principles of sustainable design and performance.

www.sustainable-homes.org.au

**Smart Housing**

The Queensland government Department of Housing's program to promote good practice in designing, planning and building homes to make them more socially, environmentally and economically sustainable.
www.smarthousing.qld.gov.au
Ph: +61 7 3238 3683
Fax: +61 7 3227 8610
Email: smarthousing@housing.qld.gov.au

## SA

**Planning SA**

This website has a section on Energy Efficient Housing within its Sustainability and Efficiency section. They also maintain a register of house energy assessors.
www.planning.sa.gov.au

## VIC

**Sustainable Energy Info**

An extensive and useful section on home building and design, covering everything from star ratings and passive solar access to advice on working with building professionals.
www.sustainable-energy.vic.gov.au

**Building Commission Victoria**

A great information resource for both consumers and building industry professionals.
www.buildingcommission.com.au

### WA

**SEDO Energy Smart Homes**

The website has an extensive section devoted to their Energy Smart Homes initiative, which includes design advice.

www.sedo.energy.wa.gov.au

## NATIONAL (NEW ZEALAND)

**New Zealand Green Building Council (NZGBC)**

A coalition of property and building industry representatives and government, aimed at reducing the harmful environmental impacts of the construction and use of buildings. At the time of publication NZGBC was progressing through an establishment phase. Over the next few years, the organisation will focus on a few key projects, including the establishment of a green home-rating system for New Zealand, a web-based hub of information on sustainable buildings, and the development of education programs.

www.nzgbc.org.nz

## Building and interior materials

**Good Environmental Choice Australia**

(formerly the Australian Environmental Labelling Association)

The Australian Ecolabel Program is an independent environmental verification and certification program that aligns with the ISO 14024 International Standards for environmental labelling. The program awards a mark

for recognition (the 'Good Environmental Choice') for products that meet standards of environmental, quality and social performance. There are a number of building and renovating products among those certified.
www.goodenvironmentalchoice.org.au

**Ecospecifier**

An online database of environmentally preferred building and interior materials. The database was produced to help architects, designers, builders and specifiers find and use materials that are sustainable.
www.ecospecifier.org
Ph: 1300 66 99 97
Ph: +61 7 3875 4173 (from outside Australia)

**Ecobalance**

A sustainable design consultancy.
www.ecobalance.com.au
Add: Unit 9b, 10 Carrington Road, Marrickville, NSW 2204
Postal address: PO Box 93, Croydon Park, NSW 2133
Ph: +61 2 9554 6655
Fax: + 61 2 9554 6688
Email: info@ecobalance.com.au

**Eco Flooring Company**

Suppliers of bamboo flooring products.
www.ecoflooring.com.au
Ph: +61 2 9402 6050
Fax: +61 2 9144 3768
Email: btbamboo@ecoflooring.com.au

**The Second Hand Building Centre**
For a range of recycled building materials.
www.shbc.com.au
Add: 432B West Botany St, Rockdale, NSW 2216
Ph: +61 2 3567 1322
Fax: +61 2 9597 1782

## Sustainable timber

**Forest Stewardship Council (FSC)**
An international not-for-profit organisation dedicated to finding solutions to the problems created by bad forestry practices and to reward good forest management.
www.fscaustralia.org

**WWF Australia—Land and Forests program**
WWF are working to ensure the conservation of Australia's landscapes and forests. They are also active in the Australian Forest and Trade Network (AFTN), the local branch of the Global Forest and Trade Network. The AFTN newsletter provides regular updates on developments in forest certification and the supply, specification and use of certified forest products.
wwf.org.au/ourwork/land/